Anonymous

Ottawa - For tourists and sportsmen

Anonymous

Ottawa - For tourists and sportsmen

ISBN/EAN: 9783742898012

Manufactured in Europe, USA, Canada, Australia, Japa

Cover: Foto ©berggeist007 / pixelio.de

Manufactured and distributed by brebook publishing software
(www.brebook.com)

Anonymous

Ottawa - For tourists and sportsmen

THE RUSSELL

OTTAWA.

F. X. ST. JACQUES

PROPRIETOR.

HOTEL VICTORIA

AYLMER, QUE.

*P*ERCHED high on a rocky promontory jutting out into the mighty Ottawa River a few hundred yards below the great Chaudiere Falls, and half a mile or so above the junction of the Gatineau with the Ottawa, stand the Parliament Buildings of the Dominion of Canada, on one of the grandest sites for public buildings to be found in the whole world. From the top of the main tower of the Parliament Buildings the view is unequalled. For a mingling of the wildness of nature and the stateliness of advanced civilization, the scenes lying before the visitor, whether he look north, south, east, or west, cannot be excelled. Looking to the north, the City of Hull, with its 15,000 inhabitants, is in the foreground, and the Chelsea Hills and the Laurentian Mountains in the background, while east and west stretches the magnificent Ottawa River, broken by the tumultious Chaudiere Falls, and widening out in the distance into the broad expanse of Lake Deschenes. A slight turn towards the east brings two more rivers into view—the Rideau, with its silvery falls, some 50 feet in height, forming a misty veil as it falls into the Ottawa on the Ontario side of the river, and the Gatineau quietly mingling its waters with the larger stream a short distance below on the Quebec side of the Ottawa.

Out of the Forest.

Facing southward, a large, handsomely-built, populous, and growing modern city lies stretched out in a magnificent panorama extending east, west, and south ; its stately churches, handsome stores and buildings and magnificent private residences half hidden in young forests of shade trees, which are

RUSSELL HOUSE.

POST OFFICE.

PARLIAMENT BUILDINGS.

growing up along all of the principal residential streets. Ottawa is essentially a city of shade. Less than a hundred years ago the site of the present city was a pine forest. The original growth has long passed away, but as the city has grown with a strong, vigorous, steady growth in the last half century, and especially within the last 15 or 20 years, the builders of the city have wisely adorned their streets with shade trees, the maple,

PARLIAMENT BUILDINGS.

chestnut, ash, rowan, and other varieties taking the place of the stately pines. In the older residential portions of the city, such as Daly Avenue, Metcale, O'Connor, and contiguous streets, there is no city in Canada where more grateful shade can be found on a hot summer's day than in the capital of the Dominion, while all the more modern portions of the city have been laid out with a view to shade in the future.

A Modern City.

Ottawa is essentially a modern city. Its foundations were laid little more than 75 years ago, when the building of the Rideau Canal was commenced under Lieut.-Col. By, of the Royal Engineers, in the year 1826, when the little settlement which grew up around the canal works as they progressed got to be called Bytown, a name which it retained until it was transformed into a city, under the name of Ottawa, by an Act of the Parliament of the old province of Canada, which took effect on the 1st of January, 1855. At that time the population was about 10,000, and the city consisted almost entirely of a collection of wooden houses on the low lying land between the line of the Rideau Canal and the junction of the Rideau River with the Ottawa, and which is now comprised in By, Ottawa, and St. George's wards of the modern city. West of the Canal there were a few stone houses and a few streets laid out, but without any buildings on them except a few wooden houses scarcely much more than shanties. At this time there was great rivalry between the cities of Quebec and Montreal in what was then known as Lower Canada, now the Province of Quebec, and Toronto and Kingston, in what is now known as the Province of Ontario, as to which city should be the seat of government of the Province of Canada, which had been created by the Union Act of 1841. A sort of perambulating capital had been established by which the seat of government was changed every four years from Toronto, in Upper Canada, to Quebec, in Lower Canada.

Selected by the Queen.

This system was found so awkward and inconvenient that in 1857 it was decided to leave the question of the selection of the site for a permanent capital to Her Majesty the Queen and she, on the advice of Her Ministers, and with the consent and ratification of the Parliament of Canada selected Ottawa. A wiser decision could not have been made. Nature appears to have designed the place for a great city, and man has not been

slow in developing the magnificent resources which have been scattered broadcast and lavishly around the Capital of the Dominion. With enormous water power rolling past her very doors and a much greater power everywhere around her, east, west, north and south ; with minerals of all kinds including iron, mica, phosphate, gold and samples at least of most of the minerals found in Canada lying hidden in the mountains which surround her ; with a vast and fertile agricultural country spreading out on every hand and reaching over to the St. Law-

CHAUDIERE FALLS.

rence, about 100 miles distant ; with half a dozen main lines of railway, and as many branch lines centering in the heart of the City ; with a vigorous and progressive people, well laid out and well lighted streets, handsome buildings, one hotel which is not excelled in the Dominion, and several others which are good in their way, a magnificent water service and the best street railway system on the Continent, Ottawa is not only a handsome, comfortable and convenient place to live in, but it undoubtedly has before it the greatest future of any City in the Dominion.

In less than half a century of corporate existence it has risen from a backwoods' village to the proud position of the fourth City in the Dominion in population, wealth, trade and importance, and there is but little doubt that ere another half century rolls away it will not only have become the third City in the Dominion but may possibly have passed Toronto and taken a position amongst the cities of Canada second only to Montreal.

Growth of the City.

Ottawa was selected as the permanent seat of Government by Her Majesty in 1857. In 1859 the public buildings required to accommodate the Parliament of Canada and the Civil Service were commenced. In 1860 the corner stone of the Parliament Buildings was laid in the basement of the Senate wing by His Royal Highness the Prince of Wales. This corner stone is one of the sights of the capital and during the session pious pilgrimages are daily made past it by Senators, Members of Parliament and visitors, which custom may, however, be partially explained by the fact that the refreshment rooms of the Senate lie immediately beyond the corner stone. In 1866 Parliament met in Ottawa for the first time. This session was the last one of the Parliament of the Province of Canada as the Confederation of all the British North America Provinces had been agreed on and the terms of the Union were then being settled. This Confederation was accomplished on the 1st of July, 1867, and on 6th of November, 1867, the first session of the Parliament of the Dominion of Canada was opened in Ottawa. At that time the population of the City was about 15,000. In 1871, according to the census of the Dominion it was 21,541 ; in 1881, 31,307 ; in 1891, 44,154. This was the last census of the Dominion, but the annual census taken by the assessors of the City show that its growth has been vastly accelerated in the past seven years, the figures given for the year 1898, being 55,386. If the present rate of growth is maintained, Ottawa will be a City of 65,000 to 70,000 inhabitants when the next decennial census is taken in 1901. This does not include the City of Hull, with its population of 15,000, or the half dozen suburbs which are almost part of the City already and which have a population of 10,000 or

15,000 more. No city in the Dominion has grown more steadily, more rapidly, and more solidly than has the City of Ottawa which now ranks as second in the Province of Ontario. Nor has this development been in population alone. Indeed, in point of trade and commercial expansion and importance the development and increase of Ottawa have been even greater than in the matter of population.

RIDEAU FALLS.

Marvellous Development.

In no city has there been greater activity in building during the last 10 or 15 years than in Ottawa; and nowhere has there been anything like the increase in railway development. In 1867 when it became the Capital of the Dominion, Ottawa could only be reached by one insignificant railway connecting with the Grand Trunk Railway at Prescott. Now some 60 passenger trains a day arrive and depart; it is the headquarters of four railway systems, two transcontinental trains pass through it every day, and almost every year sees a new line opened which either has Ottawa as its terminus, or on which Ottawa is one of the most important stations. It is the home of one of the largest and most successful financial institutions in the Dominion, the Bank of Ottawa, and no less than ten other banks have branches in the city. The assessment rolls and the customs returns show enormous increases in the last 30 years. In 1867, the assessment for the City was $5,011,840; in 1871, it was $5,970,117; in 1881, it was $10,392,275; in 1891, it was $17,638,110; and for the current year (1899) it is $23,679,275. It will thus be seen that the assessment of the City has increased 472 per cent. since Confederation. The trade of the City has

MID-WINTER SCENE.

increased in proportion to its population and increase of assessment. For the year 1867-68 the exports were $174,539. the imports $879,712, and the duty collected $79,493; for the year ended 30th June, 1898, the exports were $2,497,263; the imports $2,486,526, and the duty collected $447,982. This is an increase of nearly 1,450 per cent. in the exports, of 283 per cent. in the imports and 563 per cent. in the duty collected as compared with the figures of 30 years ago. Truly this is a phenomenal result and one of which any city would have reason to be proud.

Ottawa's Great Hotel.

The development of a city depends largely on the quantity and quality of its hotel accommodation. No modern city can hope to compete with its elder brethren until the old-fashioned "inn" or "tavern" has been replaced by a modern up-to-date hotel, thoroughly well appointed and conducted in a manner which will make a visitor from any of the large cities of the old or new world feel that he has not passed beyond the bounds of civilization when he reaches the young city. In its Bytown days, and in its earlier stages after Confederation, Ottawa was sadly deficient in the matter of hotel accommodation. Almost all the members of the first Parliament of the Dominion had to "board around," for there was not only "no room for them in the inn," but there was practically no inn for there to be room in. There were a few, very few, inns and taverns, but they were by no means of a high order of merit. At the corner of Sparks and Elgin streets there was an inn of this sort, which was about the best of its kind. Immediately after Confederation, when the Government had moved up to Ottawa, the name of this incipient hotel was changed to "The Russell House," and since then it has grown with the growth of the city, and developed with the development of the city, more than keeping pace with the city's progress, until to-day there is not a more widely known hotel throughout the length and breadth of the Dominion than "The Russell"; few which at all compare with it in size, and none which excell it in perfection of appointments, or convenience of arrangement, excellence of cuisine, or in the thousand and one little accessories which go to make up that wonderful agglomeration of comfort and convenience known as a modern hotel.

Grew with the City.

Starting from small and humble beginnings, the Russell has expanded until it occupies practically the whole block bounded by Sparks, Elgin, Queen and Canal streets, covering, with its adjunct, the Russell Theatre, very nearly an acre of

RIDEAU LAKES

ground. At Confederation, the Russell House was an unpretentious two and a half storey building at the corner of Sparks and Elgin streets. Two or three years later a stone extension on Elgin street was built. In this wing is located the magnificent dining room, 80 feet by 40, and 20 feet high, which has been the scene of almost every grand entertainment which has taken place in Ottawa. The wing is of stone, five stories high with mansard roof, the upper floors being used as bedrooms. About ten years later the eastern wing, extending along Canal street, was built, and these wings were connected by several small old-fashioned buildings which included the original hotel.

ON THE CANAL.

In 1881 all these small buildings were removed and the central portion of the hotel was built, and the name changed to " The Russell." The front of the hotel now occupies the entire block on Sparks street from Elgin to Canal streets, about 250 feet. The front elevation is five storeys with mansard roof, and is built of white brick with stone dressings. The facade is broken by a square tower in the centre and demi-towers rising slightly above the sky line at the ends of the building, the whole presenting an exceedingly solid, handsome and attractive appearance. On the Canal street front there has recently been built a

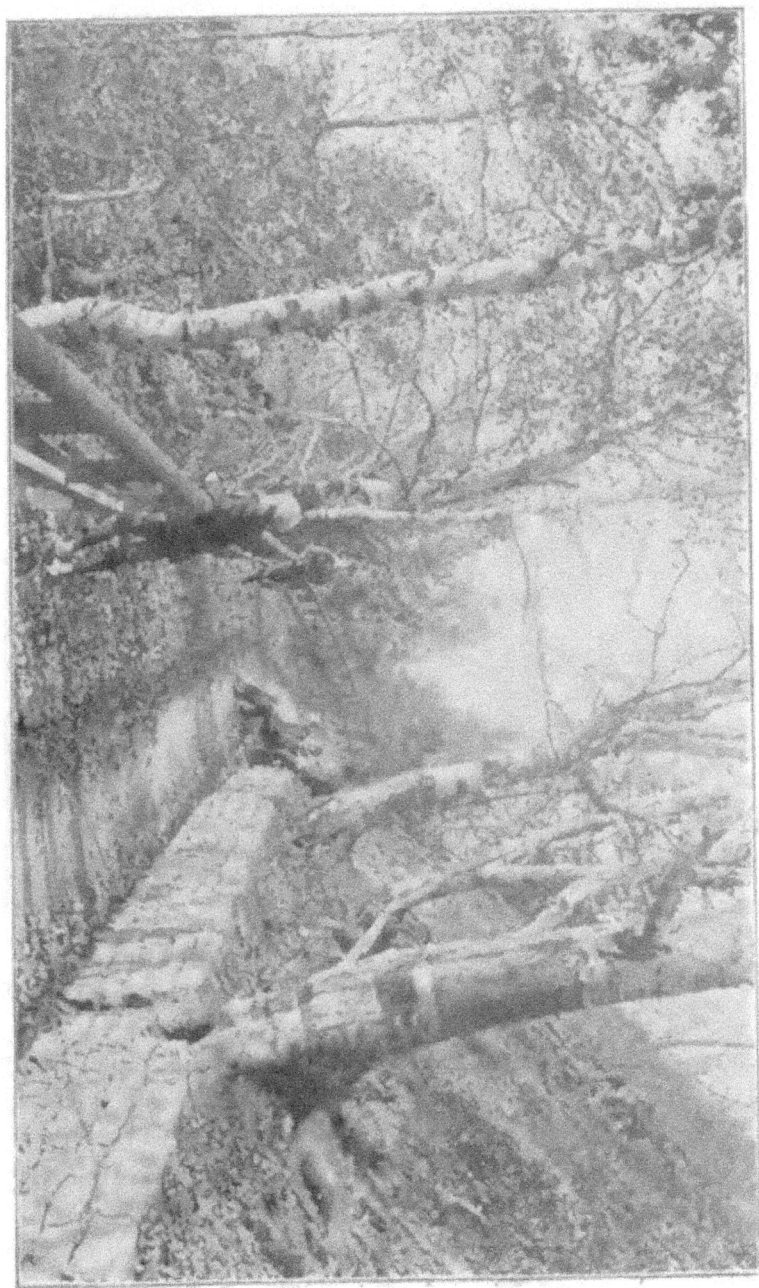

very substantial brick structure expressly designed for the accommodation of commercial travellers. This building contains 38 of the largest and most canvenient sample rooms to be found in any hotel in Canada. They are heated by steam, splendidly lighted by electricity, and both hot and cold water are laid on in each room. The building is connected with the rotunda of the hotel, and it is quite safe to say that no more commodious or convenient sample rooms can be found anywhere.

Political Centre of Canada.

When the hotel was rebuilt in 1881 its dominating feature on the ground floor was the rotunda, a magnificent room 80 by 100 feet, and this room has, to a very large extent, become the great political and commercial rendezvous of the city. Ottawa is the political centre of the Dominion. The rotunda of the Russell is the political centre of Ottawa. Here on any evening

MAJOR'S HILL PARK.

during the Session of Parliament can be seen a majority of the greater and lesser political lights who have assembled in Ottawa to perform their parliamentary duties, together with numbers of the leading political and commercial men of the Dominion, who are attracted to the capital during the meeting of Parliament either by business or pleasure, or a combination of both. Since the rebuilding of The Russell the old fashion of "boarding around" has passed out of the parliamentary mind, and now nearly one-half of the Senators and Members of Parliament take rooms at The Russell for the Session, and the rotunda and the spacious parlors on the first floor become practically adjuncts of the Houses of Parliament, located only a few hundred yards away. Then The Russell is at its best. All or nearly all of its 400 bedrooms are occupied, and its large dining room is taxed to its utmost capacity to accommodate the numbers living in the hotel, or who have rooms in the neighborhood, but take their meals at The Russell.

Thoroughly Renovated.

In the 18 years which have elapsed since the rebuilding the hotel naturally became somewhat passe, and last year the enterprising proprietor, Mr. F. X. St. Jacques, practically reconstructed, renovated, repainted and refurnished the house throughout, adding some extensions and making numerous very valuable improvements. A considerable extension was built on the Canal street front, the ground floor of which is occupied by one of the most spacious and best appointed billiard rooms to be found in any hotel. Above this are the bar and an excellent cafe with entrances into the rotunda. The rotunda itself was very greatly improved, the walls being retinted, the dome and sky-light handsomely frescoed, smoking room added and the old bar room facing on Sparks street reconstructed out of all knowledge and converted into a luxurious and handsomely appointed reading and writing room, furnished with well appointed oak desks, large oak arm chairs and a number of most invitingly comfortable lounges, while a thick and richly patterned Axminster carpet renders the room warm and comfortable to the feet. The hotel is heated thoroughly with steam and lighted by electricity.

New Fittings and Furnishings.

The finishing and furnishing is of the best and the appointments throughout the whole house from top to bottom are first-class in every respect. The dining room has also been thoroughly renovated, refitted, and the crockery replaced by French china and porcelain. There is now being specially

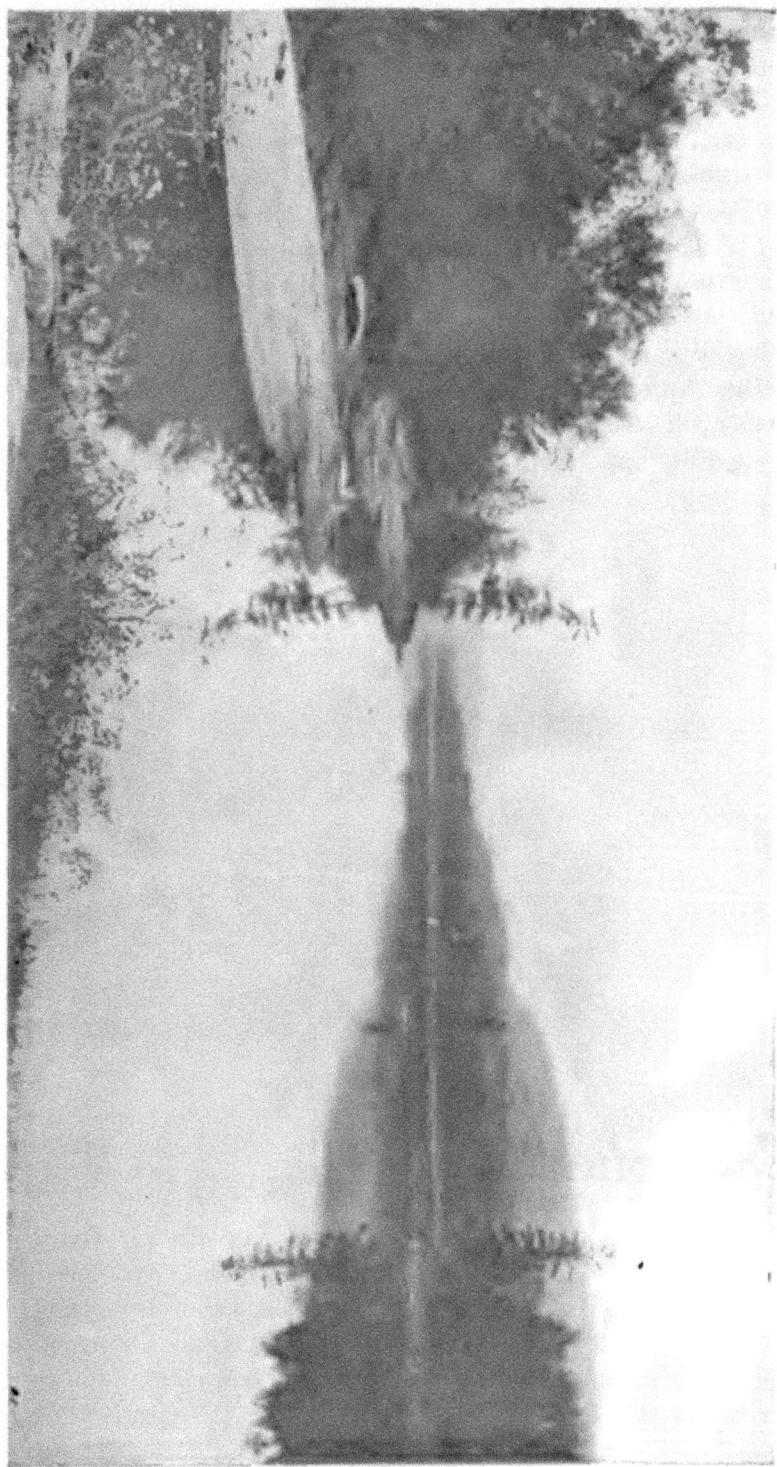

manufactured for this room a very heavy Axminster carpet, which will be one of the best to be found in any hotel and which it is expected will be completed and laid before the opening of the next session of Parliament. This dining room has a large historic interest to the whole Dominion of Canada, it having been the scene of many important gatherings, and in it every Governor General of Canada has been entertained at some time during his term of office. The improvements and alterations throughout the hotel last summer were of the most complete and exhaustive nature and nothing which could in any way conduce to the comfort of the guests of the hotel was neglected, so that to-day the Russell stands as one of the most completely and thoroughly appointed hotels to be found in the Dominion.

ON THE OTTAWA.

Only one " Russell."

In Ottawa it stands in a class entirely alone. There are many good hotels in the Capital of Canada, and some bad ones ; but there is only one Russell. This is, no doubt, largely due to the proprietor, Mr. F. X. St. Jacques, than whom there is not a better known, better liked, or more competent man in the business either in or out of the Dominion.

The Russell Theatre.

So large and prominent a hotel as the Russell would be incomplete in the modern idea of what a hotel should be, had it not two adjuncts in the shape of a theatre, or other place of amusement, and a summer resort within easy distance and under control of its management. Both these accessories are now possessed by The Russell. On the 15th of October, 1898, the Russell Theatre, the private entrance to which is from the hotel, was opened under rather auspicious circumstances in the presence of His Excellency the Governor General, Lord Aberdeen, and has since enjoyed a successful career. The Theatre is without question the best equipped and most comfortable place of amusement in the Dominion. It has a seating capacity of about 1,800 ; the auditorium is beautifully painted, frescoed and upholstered in rich dark red plush giving it an exceedingly warm and comfortable appearance, especially in the winter time. The lighting and heating are excellent. The stage and the dressing rooms accommodation are said to be more complete than in any other theatre in Canada, and the appointments throughout are first class in every respect.

The Hotel Victoria, Aylmer,

As to the summer resort, it will be difficult to find a more charming location than that of the Hotel Victoria, Aylmer, nine miles from Ottawa, reached in 20 minutes by the very efficient Electric Car service of the Ottawa and Hull Companies. This hotel, situated on the beautiful expanse of Lake Deschenes, has been recently leased for five years by Mr. St. Jacques and is being thoroughly renovated for the approaching summer season.

HOTEL VICTORIA.

Ottawa is singularly fortunate in its situation in that it appears to be within easy distance of almost any region which one might want to visit for sport, relaxation, or rest during the summer months. Taking the Russell as a centre one can in 20 minutes be transplanted to Lake Deschenes, as pleasant a place as can be visited in the summer; or, in a few hours reach some of the best fishing and hunting districts to be found in the Dominion of Canada.

Neither amusement nor rest should be the single desideratum with the man or woman to whom the consideration of the scene of the summer vacation is a subject to be deliberated upon. At this end-of the-century period it is everybody's desire to be informed, to have travelled, to know of happenings and conditions foreign to those of his or her daily life. The tourist who visits the conventional watering-place or summer resort may return home with brighter eye, more elastic step, and sun-browned countenance, but he has met no new types ; he has associated with men and women who are replicas of himself, who have the same information, the same ideas, who live during

the greater part of the year just as he does. He enjoys the change of scene ; why should he not enjoy a sojourn amongst people of another nation and of different social habits, while daily associating with those of his own ?

Not many years ago French Canada was a *terra incognita* to the great proportion of Americans and Canadians of English extraction. Latterly it has been the custom for a small number of tourists annually to make a hurried trip down the St. Lawrence ; but it was not until last year that the erection and opening of the palatial Hotel Victoria, at the picturesque village of Aylmer, Quebec, rendered it possible for the seeker after rest,

health, and amusement to see Jean Baptiste, the *habitant*, at his home, while being within half an hour's ride of a beautiful and modern city.

Nine miles from Ottawa, Ont., on the shores of the romantic Lake Deschenes—an enlargement of the majectic Ottawa—is situated the Village of Aylmer. Two hundred years ago the hardy French adventurers, " clad in doublet and hose, and boots of Cordovan leather," built many a camp-fire on the green shores of the smiling lake as they rested on their long canoe-journeys from the fort of Ville Marie (now Montreal) to the Georgian Bay, and the unknown west. As Parkman tells us, Frontenac, Champlain, Father Hennepin, La Salle ; all the adventurous Frenchmen who were the first explorers, passed the site of the Hotel Victoria.

After the conquest of Canada by the British, the population of the country rapidly increased, the French moving west, and the English settling in Ontario. To-day it is along the Ottawa river that we find the land where races blend.

Here lives the French Canadian " habitant," with his constant good humor, his scrupulously clean cottage, his parish priest—sole arbiter in all things. Not far from Aylmer—an English-Canadian village itself—the visitor will find many a cottage which might have been transferred from Normandy.

28

Providence never set a more beautiful scene than that visible from the 120-foot high observatory of the Hotel Victoria. To the north, the magnificent rock-ribbed "grand old Laurentides, old when the earth was new," lift their towering heights against an azure sky that Naples could not surpass. A stone's throw from the hotel piazza, Lake Deschenes, thirty miles in length and nine in width, sparkles in the summer sunshine. Ten miles to the east, but standing out boldly against the horizon, tower the pinnacles of the Canadian Parliament buildings, set on the magnificent limestone cliff whose steep scarp rises for two hundred feet and more from the bosom of the broad

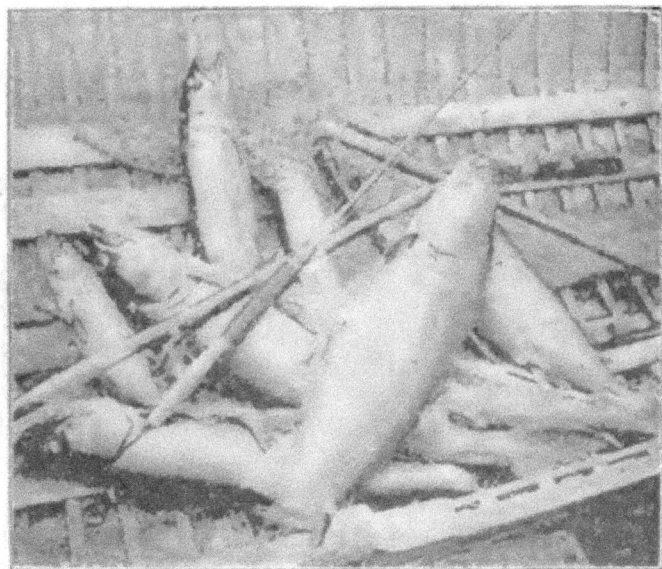

Ottawa. In the middle distance rises the column of spray from the waters which make their headlong plunge over the rocky ledge of the Chaudiere Falls, whose deep bass diapason is borne to the watcher's ear by the gentle summer wind. For three miles above these famous falls, now harnessed by, and the useful servants of, man, the Ottawa is broken into white-capped rapids, excelling in magnitude those of Niagara and impassable by craft of any kind.

Nature has devised no more perfect site for a summer resort; the Company that controls the Hotel Victoria has fur-

nished a gem that is worthy of its setting.   Environed on three
sides by a magnificent forest of ancient pines, with their health-
ful balsamic odors, with charming walks along the lake shore,
and with bathing in the softest, purest water on the continent,
the guest has all the primary conditions of well-being.   There
are other comforts, those which art and science contribute.
Modern passenger elevators, baths, 160 beautifully furnished
bedrooms, single and *en suite*, steam heating—seldom necessary
however—and the latest sanitary plumbing, make life at the
Hotel Victoria sybaritic in all but a primeval sense.   Around
three sides of the building runs a spacious piazza, twenty feet
wide, and containing 8,000 square feet of promenade.   An
amusement hall, a water-toboggan, which is so constructed

that it is absolutely safe, and an ingeniously-devised Moorish
maze, are interesting and amusing features of the establishment.   For the athletically inclined, bowling alleys, a billiard
room with new tables, and tennis courts, give opportunity for
the cultivation of the sound body which is necessary to a
healthy mind.   On Lake Deschenes a well-appointed passenger
steamer makes daily trips from Aylmer to the head of the lake,
while those preferring to be their own navigators can secure
row or sail-boats, in charge of competent men, if required.   An
orchestra furnishes music throughout the season, the hops in
the spacious ballroom being a pleasant feature, greatly enjoyed

by the guests and the *élite* of the Canadian capital. The culinary department is under the management of an experienced *chef*, and the cuisine is on the most liberal plan.

The rates are $2 per day, or $10 per week. The hotel's porters meet all trains at Ottawa, whence guests are transferred to the doors of the Hotel Victoria in twenty minutes in the palatial cars of the Ottawa, and Hull & Aylmer Electric Railways.

From Aylmer the guest can indulge in many side trips. Ottawa, with its excellent stores and asphalted streets, is within half an hour's run. In Ottawa the visitor can profitably

spend a few days at the far-famed Russell House, an establishment known all over the continent as being " the Palace Hotel of Canada. Under the skilful management of Mr. F. X. St. Jacques, the proprietor, the best known hotel man in Canada, the Russell has achieved a well merited reputation. As the accepted headquarters for all the distinguished statesmen, travellers, and kings of commerce who visit Ottawa, its registers for many years bear the autographs of men whose fame is world wide. Containing four hundred elegant bedrooms, with all the modern adjuncts, the Russell is always patronized by those who know what hotel comfort is. Within the past few months improvements costing many thousands of dollars have been made, until at the present time the proprietor of The Russell can justly claim to have an establishment which will compare favorably with any hotel anywhere to be found. Situated near the grand Parliament buildings, the Russell furnishes convenient headquarters for visitors having business with any of the Government departments.

For the cyclist, there are several ideal runs ; to Chelsea, with its beautiful waterfall ; and across the lake to the pretty villages of Britannia and South March. Montreal, the commercial metropolis of Canada, is only 3½ hours distant. The

new Pontiac and Pacific Railway also gives the sojourner an opportunity of spending a day in what, two years ago, was a primeval forest, and in which to-day the moose, red deer, wolf, and beaver are plentiful.

Take it all in all, there is no summer resort on the continent which presents so many and varied advantages. The success of the Hotel Victoria last season was a proof that in 1899 and future years, it will be one of the most prominent summering places in North America. Its accessibility, its beautiful situation, the determination of its proprietors to make it a genuine *hotel de luxe*, are all earnests of the inevitable and continued popularity of the Hotel Victoria, "where races blend,"

Sportsman's Paradise.

Just across the Ottawa river, and easily reached by the Gatineau Valley Railway, is the great fishing region of the Laurentian Mountains. The Gatineau country is fairly honeycombed with thousands of lakes, large and small, which are fairly teeming with fish. Black bass, grey and speckled trout, maskinonge, and other varieties of game fish, fairly abound, and although there are game laws in the Province of Quebec, and a large proportion of the lakes are under lease for fishing privileges; still the visitor at The Russell need never fear but that he can get all the fishing he wants, as the game laws are not severe and fishing privileges are easily obtainable. The sport within easy reach of Ottawa is not confined to fishing. Along the Ottawa, Arnprior and Parry Sound Railway, which passes through Algonquin Park; on the lines of the Gatineau Valley and Pontiac Pacific Railways, which penetrate the Gatineau region and touch the Laurentian hills in the Pontiac region, there is excellent hunting. Partridge, ducks of all kinds, caribou, red deer and moose are abundant; and along the Lievre, Blanche, Rouge, Upper Gatineau, Colonge and Dumoine rivers, on the Quebec side of the Ottawa, and the Madawaska, Bonnechere, Schyoun, and Pettawawa rivers, in Ontario, most excellent sport can always be had during the autumn months.

33

Ottawa to Aylmer.

Aylmer is reached from Ottawa by the cars of the Hull Electric Railway Co., which pass the grounds of the Hotel Victoria every ten minutes.

✠ ✠ ✠

How to Reach Ottawa.

From Niagara Falls and Buffalo.

By the Michigan Central, Toronto, Hamilton & Buffalo and Canadian Pacific Railways. All rail.

By the Grand Trunk Railway to Brockville, and the Canadian Pacific Railway. All rail.

By the Grand Trunk Railway to Coteau Junction, and Canada Atlantic Railway. All rail.

Also by the Michigan Central, Toronto, Hamilton & Buffalo, and Canadian Pacific Railways to Toronto, Richelieu & Ontario Navigation Company's steamers to Prescott (passing through the Thousand Islands) and Canadian Pacific Railway.

By the Grand Trunk to Toronto or Kingston, Richelieu & Ontario Navigation Company's steamers to Prescott, and Canadian Pacific Railway; or to Coteau Landing, and Canada Atlantic Railway.

By the New York Central Railway to Lewiston, Niagara Navigation Company's steamers to Toronto, Canadian Pacific and Grand Trunk Railways, all rail; or Richelieu & Ontario Navigation Company's steamers, through the Thousand Islands, to Prescott, and Canadian Pacific Railway; or to Coteau Landing and Canada Atlantic Railway.

From Toronto.

By the Canadian Pacific or Grand Trunk Railways, all rail; or by Richelieu & Ontario Navigation Company's steamers to Prescott, and Canadian Pacific Railway; or to Coteau Landing, and Canada Atlantic Railway; or Grand Trunk Railway to Kingston, Richelieu & Ontario Navigation Company's steamers to Prescott, and Canadian Pacific Railway; or to Coteau Landing, and Canada Atlantic Railway.

From the Thousand Islands.

By regular steamers to Brockville or Prescott, and Canadian Pacific Railway.

From Kingston.

By steamer "Jas. Swift" through the far-famed Rideau Lakes, sailing from Kingston every Monday and Thursday at 6 a.m.

From Montreal and Points East.

By the Canada Atlantic Railway.

By the Canadian Pacific Railway.

By the Ottawa River Navigation Company.

From New York, Troy, Albany.

By New York Central, Fitchburg, Bennington & Rutland, Rutland, Central Vermont and Canada Atlantic Railways. (Through sleeping cars by this route both ways without change.)

By New York Central, Delaware & Hudson and Canada Atlantic Railways.

By New York Central, Adirondack & St. Lawrence and Canadian Pacific Railways.

By New York Central Railway, Rome, Watertown & Ogdensburg Ferry (Ogdensburg to Prescott), and Canadian Pacific Railway.

From Boston and New England Points.

By the Boston & Main, Central Vermont and Canada Atlantic Railways.

By the Fitchburg, Rutland, Central Vermont and Canada Atlantic Railways.

By the Boston & Maine and Canadian Pacific Railways.

From the Far-Famed Muskoka District, Parry Sound, etc.

By the Ottawa, Arnprior & Parry Sound Railway.

———

Drawing room cars on day, and sleeping cars on night trains to and from Ottawa over all lines.

Open Seasons for Game and Fish.

✠ ✠ ✠

Synopsis of laws governing shooting and fishing in the Provinces of Ontario and Quebec.

NOTE—The following condensations of the Game Laws, etc., have been carefully revised, and made as correct as possible up to the date of the issue of this pamphlet. Owing to the fact that game laws are frequently changed, absolute accuracy is not guaranteed.

PROVINCE OF ONTARIO.

SHOOTING—Moose, caribou, elk, and reindeer protected entirely until October, 1900. . . . Deer can only be hunted taken, or killed between November 1st and November 15th, i.e, 15 days. . . . Quail and wild turkeys, September 15th to December 15th. . . . Grouse, pheasants, woodcock, golden plover, prairie fowl, partridge, snipe, rail, hare, 15th September to 15th December following. . . . Swans and geese, 15th September to 1st May. . . . Ducks of all kinds and other waterfowl, 1st December to 15th December. No person shall shoot between sunset and sunrise. Cotton tail rabbits may be shot at all times.

No person can kill deer in Ontario, except he hold a license from the Provincial Secretary. No person shall kill more than TWO DEER, and deer are not to be hunted or killed in the water.

No person shall kill or take any moose, elk, reindeer, caribou, deer, partridge or quail, for the purpose of exporting the same out of Ontario.

FISHING—Open Season—Salmon, trout, and whitefish, between the 1st November and 1st December. . . Speckled trout, brook trout, river trout, from 1st May to 15th September. . . . Bass and maskinonge from 15th June to 15th April. . . . Pickerel, 15th April to 15th May. No person shall kill more than fifty speckled or brook trout in one day, or more than aggregates in weight 15 pounds, or any trout less than five inches in length. Smaller ones to be returned to the water. Not more than one dozen bass to be killed in one day, or any less than ten inches long.

HUNTING AND SHOOTING IN QUEBEC.

No person who is not domiciled in the Province of Quebec, can, at any time, hunt in this Province, without having previously obtained a license to that effect from the Commissioner of Lands, Forests and Fisheries, or by any other person designated by him. Such license is not transferable and shall be good only for the hunting or shooting season for which it is issued. The fee may be reduced if the license is issued to a member of any Fish and Game Club which is incorporated under the laws of the Province and has complied with the provisions of such law; but on condition that such Club is lessee of a hunting reserve in accordance with Article 1417a. The hunting rights does not give non-residents of the Province the privilege to fish.

N.B.—Fine of $2 to $200, or imprisonment in default of payment.

FISHING IN QUEBEC.

OPEN SEASON,

BASS—From 16th June to 15th April.

MASKINONGE—From 2nd July to 25th May.

PICKEREL—DORE—From 16th May to 15th April.

SALMON—From 2nd February to 15th August.

SPECKLED TROUT—From 1st May to 1st October.

GREY TROUT, LAKE TROUT OR LUNGE—From 2nd December to 15th October.

OUANANICHE—From 2nd December to 15th September.

WHITE FISH—From 2nd December to 10th November.

FISHING.

ARTICLE 1378—No person who is not domiciled in the Province of Quebec, can, at any time, fish in the lakes or rivers under control of the Government of this Province, not actually under lease, without having previously obtained a license to that effect from the Commissioner of Lands, Forests and Fisheries, or by any other person designated by him.

FEE FOR LICENSE.

The fee required is determined in each case by the Commissioner, but it shall never be less than ten dollars.

Such licenses are only valid for the time, place and persons therein indicated.

N.B.—Angling only by hand (with hook and line) is permitted, for taking fish in the lakes and rivers under the control of the Government of the Province. For any other mode of fishing a special authorization of the Commissioner is required. The fishing rights does not give non-residents of the Province the privilege to hunt.

FINES.

Five ($5) dollars to twenty ($20) dollars or imprisonment in default of payment.

TRANSPORT.

After the expiration of the ten days allowed for transportation after the close of the open season, all railway, steamboat and other companies and public carriers are forbidden to carry any kinds of fish.

Any railway, steamboat or other company, or any person favoring in any manner whatever the contravention of this article, shall be liable to a penalty of not less than two dollars and not more than twenty dollars.

Nevertheless, it is lawful for the Commissioner of Lands, Forests and Fisheries, at any time, to grant transport permits, when it has been established to his satisfaction, that the fish which it is desired to transport, have been taken during the time when fishing is allowed and in a lawful manner.

For such permits there may be exacted a fee, the amount whereof shall be fixed by the Commissioner, according to circumstances, but which shall not exceed five dollars.

Canadian Customs Regulations

In Regard to

Tourists and Sportsmen's Outfits.

✠ ✠ ✠

The articles which may be brought into Canada (in addition to wearing apparel, on which no duty is levied), as tourist outfifs, comprise guns, fishing rods, canoes, tents, camp equipment, cooking utensils, musical instruments, kodaks, etc.

A deposit of duty on the appraised value of the articles imported must be made with the nearest Collector on arrival in Canada, which deposit will be returned in full, provided the articles are exported from Canada within six months.

REPORT FOR DEPOSIT ON TOURIST'S OUTFIT.

(In Duplicate)

Entry No................... Report No. ....

Port of...

.............................189....

Tourist's Outfit imported by ....

of........................... per...........................

from.......................................

Marks and Nos.	Descript'ons of Articles.	Value.	Rate.	Duty.	Remarks re Exportation.

The said deposit of..................... ....dollars has been received by me on the conditions stated by the importer.

....
Customs Officer.

```
STAMP.
```

I...(owner or agent) do
solemnly declare that the above is a full and true statement of
the description and values of the articles imported by me as
Tourist's Outfit, with the amount of duty deposited thereon, the
said deposit to be entered for duty if the articles are not duly
exported within six months.

(Signature)...................................

(Address) ....

INSTRUCTIONS.

If the tourist is unable to have his outfit exported and
identified at the Customs port where the deposit of duty is
made, so as to receive back his deposit before leaving Canada,
he can have the articles inspected and certified as below. The
Tourist's Report of the articles exported and certified as aforesaid may then be mailed to the Customs Officer at the port of
entry, who will forward a remittance, by mail, for the money
deposited (less expense of remittance).

The articles which may be brought in as Tourists' Outfits
comprise : Guns, Fishing Rods, Canoes, Tents, Camp Equipment, Cooking Utensils, Musical Instruments, Kodaks, etc.

*Declaration as to return of Outfit, attested to before a Customs
Officer in Canada or at a place out of Canada.*

Articles described herein inspected by me at.............
this.....................day of ....18....
and exported or landed as declared.

Sworn to before me,

....

Customs Officer.

STAMP

I....................................(owner or agent) do
solemnly declare that the identical goods hereinbefore described
are now presented for inspection, the same having been de-
livered for exportation from the port of ....
or landed at......................from.......... ....
per...................................

(Signature)

....

42

Club Visitors' Shooting and Fishing Outfits.

✠ ✠

MEMORANDUM.

CUSTOMS DEPARTMENT,
OTTAWA, July 15th, 1898.

To Collectors of Customs:

1. Any organized Shooting or Fishing Club, which has duly obtained shooting or fishing privileges in any Province in Canada, may deposit with the Department of Customs at Ottawa, a Guarantee, as per Schedule "A" hereto, as security for the due exportation or payment of duty on the sporting outfits brought into Canada temporarily, for their own use and not for gain or hire, by members of the Club resident outside of Canada.

2. The "Guarantee," if approved by the Minister of Customs, shall continue in force for the time specified therein, unless otherwise ordered, and "Guarantee Certificates" indicating such approval may be signed and forwarded to Shooting and Fishing Club by the Commissioner of Customs, to be presented to Customs Officers (and returned to visiting members after inspection) as evidence of the deposit and approval of the Club Guarantee.

3. A special Ticket of Membership signed by the Secretary of the Club, and dated within one year from the time of its presentation to the Customs Officer, may be accepted by such officer as evidence that the person presenting the Ticket is a visiting member of such Club, resident outside of Canada.

4. Visiting members (non-resident in Canada) of any Shooting or Fishing Club which has deposited a "Guarantee" approved by the Minister of Customs as herein provided, may bring with them such guns, fishing rods and sporting outfit as they require for their own use and not for gain or hire, conditional on exportation within ninety days from time of entrance, upon depositing a Ticket of Membership signed as before men-

tioned, and furnishing to the Customs Officer at the Port o
Entry in Canada a report (in duplicate) signed by him and containing a description of the articles comprised in his outfit, and
giving the value thereof : Provided, however, that *duty shall be
paid on ammunition and provisions* brought in with such outfit.

5. One copy of the report, stamped by the Customs Officer
for identification, shall be handed back to the party presenting
the same, so that a certificate as to the exportation of the
articles may be endorsed thereon by a Customs Officer when
they are not exported outwards and examined by a Customs
Officer at the Port of Entry where they have been brought into
Canada.

6. At the expiration of three months from time of entry,
Customs Officers shall forward the Outfit Reports to the Customs Department at Ottawa for action thereon, unless articles
have been exported under their inspection or a certificate of
such exportation signed by a Canadian or foreign Customs
Officer has been received by them or the duty paid on any article not duly exported.

7. The form (E 32) in Schedule " B " hereto may be used
for Report and Certificate of Exportation *re* " Club Visitors'
Shooting and Fishing Outfits."

8. The "Guarantee Certificate" may be in Form " C "
thereto.

JOHN McDOUGALD,
Commissioner of Customs.

"A"

GUARANTEE FOR EXPORT OF SHOOTING AND FISHING CLUB OUTFITS.

The Commissioner of Customs of Canada:

Referring to Memorandum No. 1006 B, issued by the Customs Department of Canada on the....................day of
July, 1898, the undersigned Club, known as ....
......................with headquarters at................
hereby applies for the temporary admission into Canada of the
Sporting Outfits of the members of the said Club resident outside of Canada, without payment of duty on such outfit, at the

time of its importation ; and in consideration thereof and for other considerations the said Club and......................
of................................and ....
of.......................... do jointly and severally hereby guarantee to pay to Her Majesty the Queen, all Customs duties due and to become due on the whole or any part of the outfits of the said members admitted into Canada as aforesaid, between the................day of.......................189..., and the................day of.......................189..., unless they shall be duly exported in accordance with the requirements of the said Memorandum.

Dated at...........................this..............day of........................18....

(Name of Club)..

(President or Chief Officer)............................

(Signature, Secretary) ....

In presence of

(Witness) ....

The above Guarantee is approved and filed the............
day of........................18....

....

"B"

FORM E 32.—CUSTOMS, CANADA, re CLUB VISITORS, SHOOTING AND FISHING OUTFITS.

Port........................ } Report No.................
Entry No.................

Report of Club Visitor's Outfit imported by..............
of........................per........................
from..........................;

Marks and Nos.	Description of Articles.	Value.	Rate.	Duty.	Remarks re Exportation.

I.............................(owner or agent) do
solemnly declare that the above is a full and true statement of
the description and values of the articles imported by me as a
Club Visitor's Outfit, conditional on payment of duty on all
the articles which are not duly exported within ....days.
>   (Signature) ....
>   (Address) ....
....
Customs Officer.

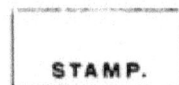

Guarantee by Club known as ....

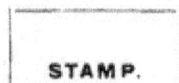

┌──────────────┐
│   │
│   STAMP.   │
│   │
└──────────────┘

*Declaration as to return of Outfit, attested to before a Customs
Officer in Canada or at a place out of Canada.*

Articles as described herein, inspected by me at ....
this..................... day of.....................18....
and exported or landed as declared.
Sworn to before me,
....
Customs Officer.

┌──────────────┐
│   │
│   STAMP.   │
│   │
└──────────────┘

I.............................(owner or agent) do
solemnly declare that the identical goods hereinbefore described
are now presented for inspection the same having ,been de-
livered for exportation from the port of.....................
per.......................or landed at ....
from...........................per ....
>   (Signature)
....

NOTE—If the tourist is unable to have his outfit exported
and identified before leaving Canada, at the port where reported
inwards, he can have the articles inspected and certified as
above. The tourist's report of the articles exported and certified
as aforesaid may then be mailed to the Customs Officer at the port
of entry, and duty must be paid on all articles not re-exported.

"C"

GUARANTEE CERTIFICATE.

The guarantee of..................... continuing in force
from..........,..................to ....
for Club Visitors Shooting and Fishing Outfits, as provided by
Memo. No. 1006 B of July 15th, 1898, has been duly deposited
and approved at the Customs Department of Canada, on the
...........................18....

....

Commissioner of Customs.

Fire Law.

✠

LESSEE OF FISH AND GAME PRESERVES

ANSWERABLE for damages to timber by waste, etc., fire, etc.,
unless he prove that all due precaution has been taken.

He shall further be answerable for damages caused by himself or by the people under his control, to the timber growing
on said territory, or on the adjoining territory, either from
waste or want of sufficient precaution in lighting, watching
over or putting out fires, and it shall be incumbent on him, in
case of damage done by fire, to prove that all such precautions
have been taken. 51-52 V., C. 17, S. 6.

THE

Best Fishing and Hunting Grounds

IN CANADA

ARE IN THE PROVINCE OF QUEBEC.

INFORMATION BUREAU.

Full information concerning hunting grounds for MOOSE, CARIBOU, RED DEER, PARTRIDGE, etc., and Fishing Lakes or Streams for SALMON, SALMON TROUT, SPECKLED TROUT, BASS, PICKEREL, etc.

Routes to Fishing and Hunting Grounds, Accommodation, Selection of Guides and Canoes, Rentals of Hunting and Fishing Preserves, Permits for Non-Residents, and all other information can be furnished on application to

N. E. CORMIER,

Provincial Game Warden and Fishery Overseer,

AYLMER EAST,

PROVINCE   QUEBEC, CANADA.

www.ingramcontent.com/pod-product-compliance
Lightning Source LLC
Chambersburg PA
CBHW022026190326
41519CB00010B/1614